HOW IT'S BUILT
SAILBOAT

by Rebecca Stanborough

Illustrations by Richard Watson

Children's Press®
An imprint of Scholastic Inc.

Thanks to Christopher Sanders, Shipwright, Mystic Seaport, for his role as content consultant for this book.

Thanks to Donna Lowich, Senior Information Specialist at the Christopher & Dana Reeve Foundation, for her insights into the daily lives of people who use wheelchairs.

Library of Congress Cataloging-in-Publication Data available
ISBN 978-1-338-80014-2 (library binding) ISBN 978-1-338-80015-9 (paperback)

10 9 8 7 6 5 4 3 2 1 22 23 24 25 26

Printed in the U.S.A. 113
First edition, 2022

Series produced by Spooky Cheetah Press
Book design by Maria Bergós, Book & Look
Page design by Kathleen Petelinsek, The Design Lab

Photos ©: cover: Ty Wright/Bloomberg/Getty Images; back cover: Elliot Elliot/age fotostock; 5 top: Sentilo Media/Alamy Images; 8 right: Roy Hulsbergen/Dreamstime; 10–11 blueprint: Jelena83/Getty Images; 11 inset: Dorling Kindersley Ltd/Alamy Images; 12 right: SafakOguz/age fotostock; 13 top left: noprati somchit/Getty Images; 13 top right: Sergiy1975/Dreamstime; 14–15: Borislav Dopudja/Alamy Images; 15 inset: Kos Picture Source Ltd/Alamy Images; 16 left: Uwe Moser/Getty Images; 16 right: Cavan Images/Alamy Images; 17 left: Cavan Images/Alamy Images; 17 right: Borislav Dopudja/Alamy Images; 18–19: Patrick Eden/Alamy Images; 21 worker: Bill Schmidt; 22 left: Onne van der Wal/Getty Images; 22 right: Matthew Mawson/Alamy Images; 23 left: Toby Roxburgh/Nature Picture Library; 23 right: Mint Images/age fotostock; 24–25: Keith Skingle/Alamy Images; 26–27: Sentilo Media/Alamy Images; 28 bottom: Cavan Images/Alamy Images; 29 top left: Sasin Tipchai/Dreamstime; 30 left: Gianni Dagli Orti/Shutterstock; 30 right: Bristol Archives/Marriott Collection/Bridgeman Images; 31 bottom: Zoonar/Elena Duverna/age fotostock.

All other photos © Shutterstock.

TABLE OF CONTENTS

• MEET THE JUNIOR ENGINEERS CLUB •

These six friends love learning about how things are built! This is their workshop.

Jacob and Nisha found out how a sailboat is built. Now they are sharing what they learned!

A small sailboat takes two weeks to build. Bigger sailboats take longer.

LET'S BUILD A SAILBOAT!

Hi! I'm Jacob, and this is my friend Nisha. We're going to a sailboat race this weekend. First we wanted to know about the boats that would be in the race. We met Raye at the marina. She is a naval architect who designs sailboats. Raye told us that all sailboats have a few parts in common.

The **mast** is the long aluminum pole that holds up the sails.

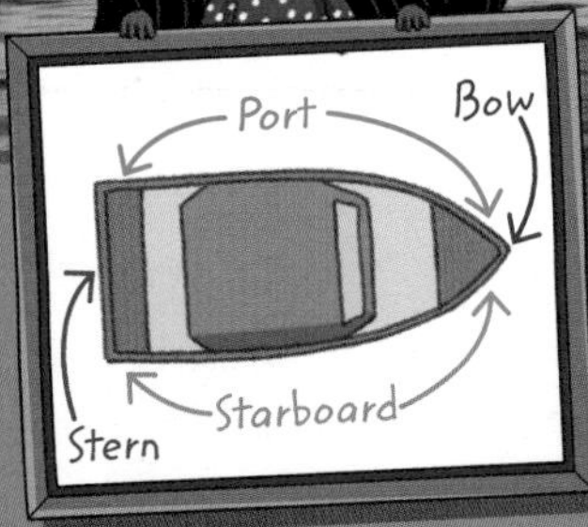

The front of a boat is called the bow. The back is the stern. The left side is called the port, and the right is called the starboard.

Sails are big pieces of cloth that catch the wind. That causes the boat to move.
All the boats in the race will be dinghies, like this one.
The **deck** is the top of the boat. That is where the sailor sits.
The **rudder** is used to steer the boat.
The **hull** is the bottom of the boat. That's the part that floats in the water.

Raye explained that the dinghy we were learning about is just one type of sailboat. There are lots of different kinds.

A **sloop** has one mast and two sails. Big sloops have cabins inside their hulls. A cabin is like a tiny house with a kitchen, bed, and bathroom.

Schooners, **ketches** (above), **yawls**, and **brigantines** all have two masts. The mainmast is in the middle of the boat. The mizzenmast is near the back of the boat.

A **catamaran** has two hulls. The hulls on a small catamaran are connected by a trampoline. The hulls on a big catamaran are connected by a bridge.

A **dinghy** is small—just 5 to 12 feet (1.5 to 3.7 meters) long. It has one mast and can be sailed by one or two people.

I went on a catamaran once. We sat on the trampoline.
The trampoline also keeps waves from flipping the boat.

Raye uses a computer when she designs a boat. Dinghies usually have a hull shaped like the letter "U" or the letter "V." The sides of the hull are curved to help the boat rock back up when a strong gust of wind tips it to one side.

Why is the front of the hull so long and narrow?
So it can cut through water like a knife!

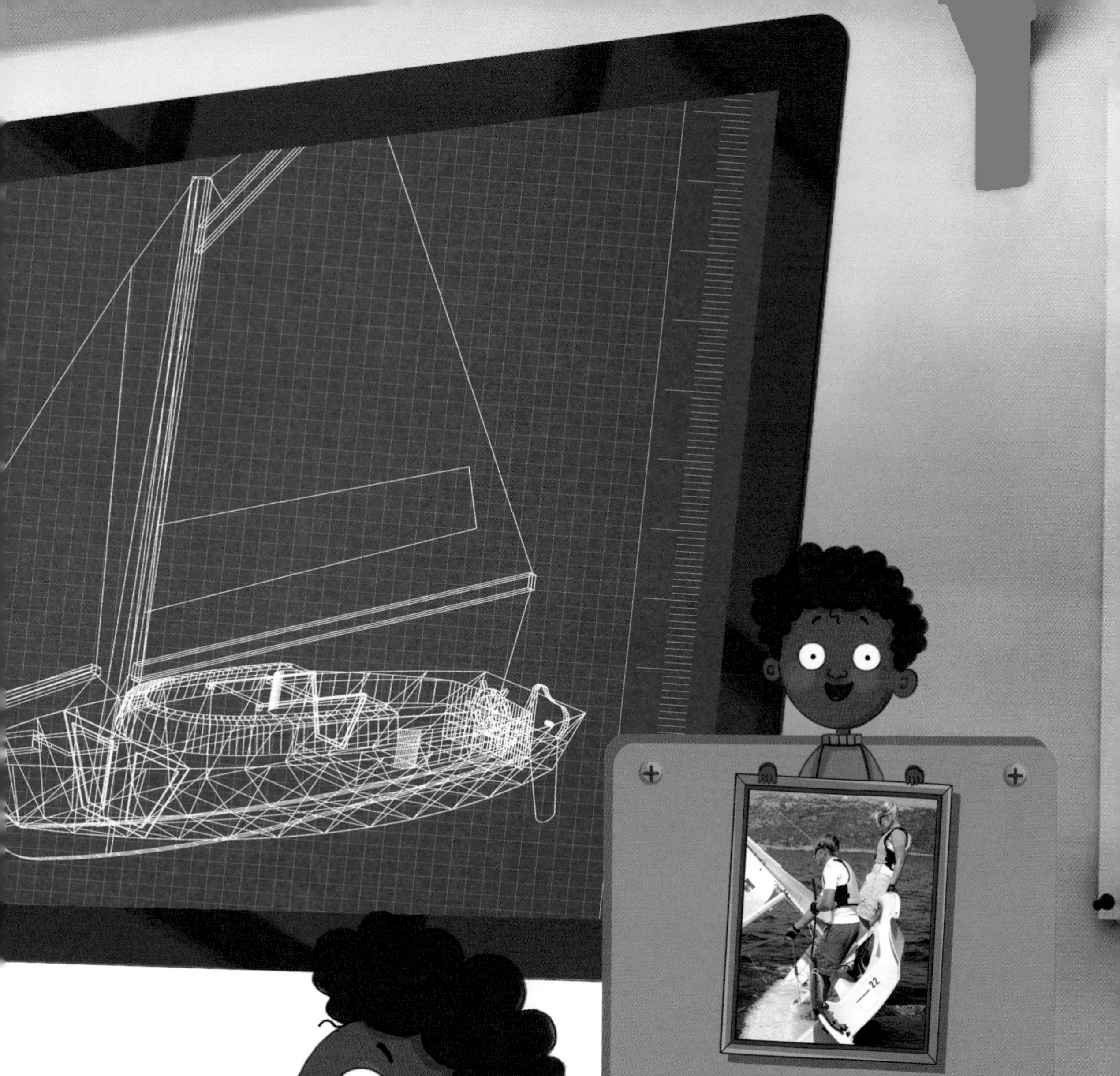

A boat's weight has to be carefully balanced. Otherwise, one part of the boat might dip too close to the water.

Raye showed us some of the materials that are used to build racing dinghies. The boats have to be fast, so they are made of light materials. But the materials also have to be strong. Sunshine, seawater, and wind can cause a lot of damage!

Fiberglass is cloth made from chopped glass. The glass is mixed with plastic and rolled into mats. Fiberglass is used to build hulls and decks.

Aluminum is a light metal that is used to build masts. The salt in seawater will not make holes in aluminum.

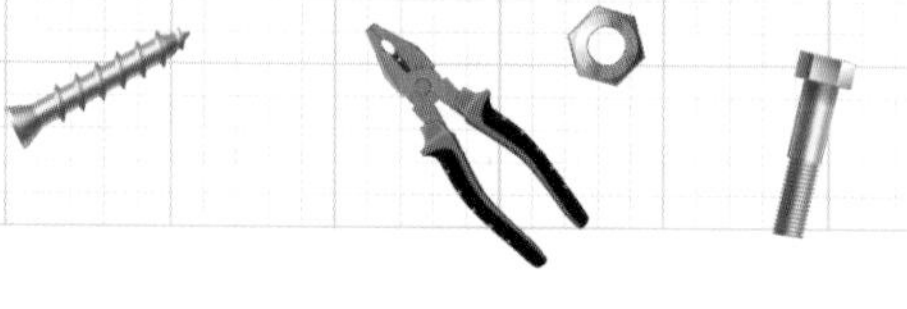

Resin is a sticky coating that is painted onto the outside of a boat. When resin dries and hardens, it keeps water from seeping into the hull and deck.

Sails are woven from a material called polyester. Polyester stretches so it can catch the wind without ripping.

Resin looks super messy.
And super sticky!

Sailors pull ropes, called **lines,** to move the sails. The lines are made of long strings of polyester that are twisted together to make them strong.

Then we met Lewis. He was the engineer in charge of building the dinghy. Lewis explained that to build a fiberglass boat, the builders first had to make two molds. One was for the hull, and one was for the deck.

This mold is for the hull.
I can see the shape of the boat!

Most sailboats today are made of fiberglass, but some are made of wood. Builders don't use a mold for wooden boats.

Once the molds were ready, the builders could make the deck and the hull. They completed the same process for each.

1

First, boatbuilders sprayed gel paint into the mold. The first layer inside the mold will ultimately be the outside of the boat!

2

After the paint dried, the builders laid fiberglass cloth inside the mold. This process is called glassing the boat.

3

Next, the first layer of cloth was covered with resin. Sometimes the resin is sprayed on. Sometimes it is painted on by hand. Then the builders repeated steps 1 and 2.

4

After the layers had hardened, that part of the boat was ready to be popped out of the mold.

It takes many layers of fiberglass cloth and resin to build a strong boat.
That's amazing! The fiberglass started out so soft!

The boatbuilders cut any extra fiberglass off the deck and hull. Then they shined the whole boat with a buffer. That process is called trimming and polishing.

Next, the builders lifted the deck over the hull and lowered it until both parts of the boat fit together. They used glue, metal screws, and metal bolts to join them.

Buffer
Trimming and polishing are not just about making the boat pretty. The boat needs to be perfectly smooth so it can move quickly through the water.
Our boat is almost finished!
I can't wait to see it when it's done!

Next, we visited the mast workshop. The mast is made from aluminum that has been shaped into a tube. Special machines are used to harden the mast so it won't break in a storm.

Most large sailboats have a wind indicator at the top of the mast. That instrument allows the sailor to determine the boat's course and how to adjust the sails.

Is the mast narrower at the top?
Yes. That's so it can bend in the wind.

After the mast workshop, Lewis and Raye showed us the loft. That's the place where sails are made. This dinghy will have just one sail.

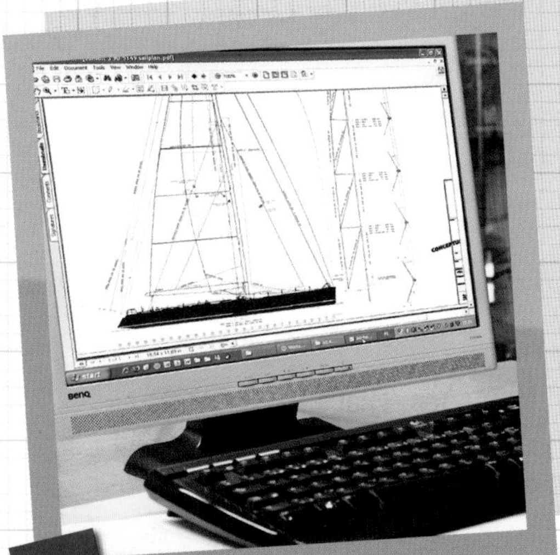

1

First, sailmakers designed the sail on a computer. The computer helps them see how a sail's shape changes in wind and rain.

2

Then the sailmakers used a machine called a plotter to draw the shape of the sail on a long piece of cloth.

3

Next, a laser cut out the sail in the right size and shape. Small dinghies have sails made from one piece of cloth. Bigger sailboats have bigger sails. Those are made of many pieces stitched together.

4

Finally, the sailmakers hammered metal rings through the cloth. Ropes and wires will run through the rings to make it easier to raise and lower the sails.

Maybe. But that step must be very precise.
Using the plotter looks like fun!

It was time for the finishing touches! First the boatbuilders attached the centerboard to the hull. The centerboard looks like a fin under the boat. It keeps the wind from pushing the boat from side to side, and folds up when it's not in use. Then they attached the rudder. The rudder is used to steer the boat. Finally, the builders attached the mast and sail.

I guess the rudder is sort of like a steering wheel.
Rudder

In a race, every dinghy must have a number on the sail and the hull. In an international race, the sail must also have letters that show the sailor's home country.

It doesn't look like one, though!

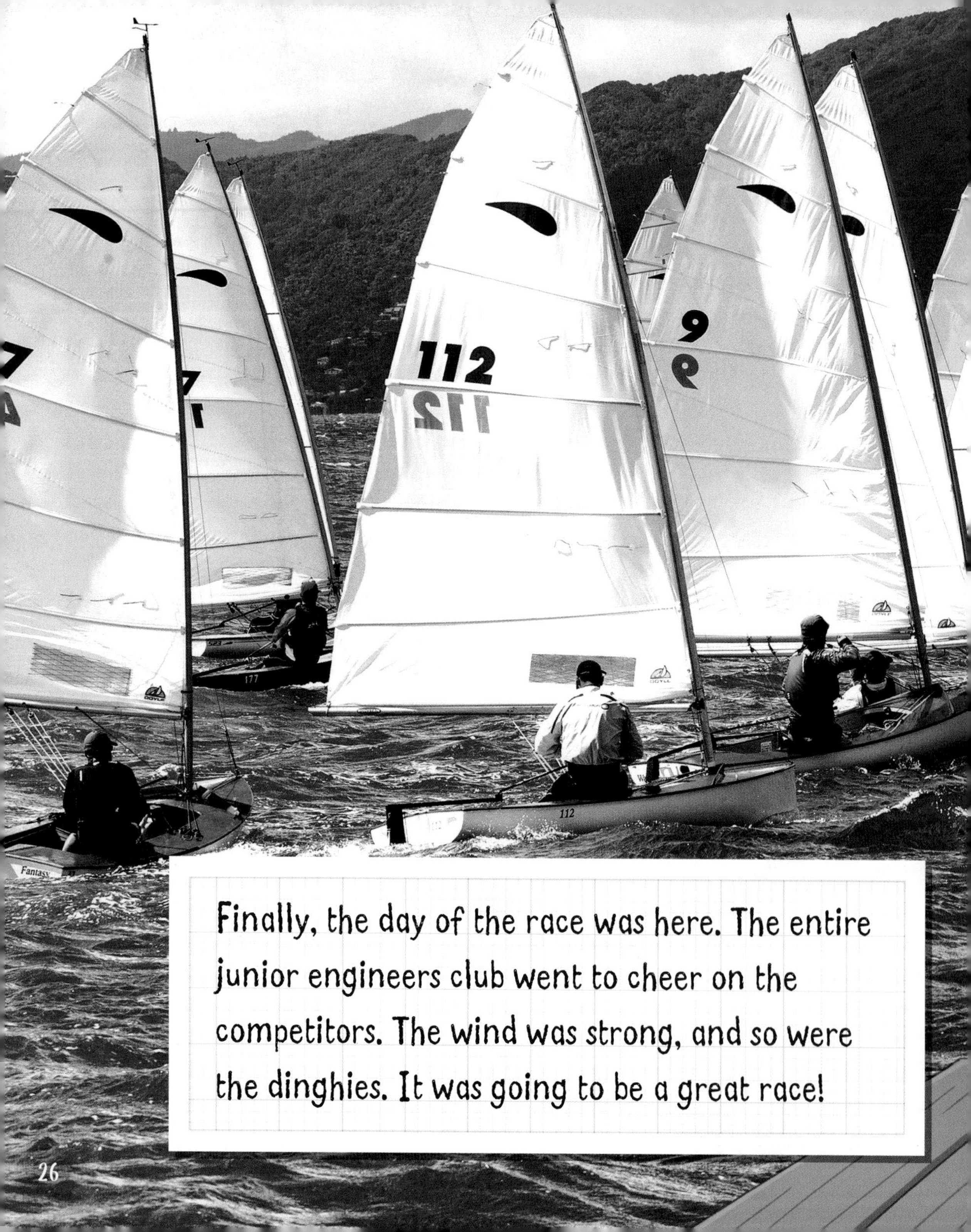

Finally, the day of the race was here. The entire junior engineers club went to cheer on the competitors. The wind was strong, and so were the dinghies. It was going to be a great race!

Yay!
Look at them go!

MACHINERY AND TOOLS FOR BUILDING A SAILBOAT

Saw
This tool is used to cut wood for large boats that have cabins or wooden decks.

Spray Gun
A spray gun has many uses. It can be used to spray resin into a mold. It can also be used to coat the mold with gel paint.

Sander
A sander uses electricity to spin gritty sandpaper. The sandpaper scrapes the splinters off wood to make it smooth.

Crane
This heavy machine lifts and moves the mold, hull, and deck.

Drill
This small tool can be used to attach hardware to the hull, deck, and mast.

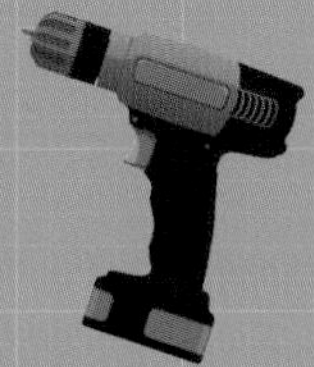

Trailer
When the boat is finished, a flat trailer with wheels carries the sailboat to the water.

THE WORLD'S FIRST SAILBOATS

People have been building sailboats for more than 6,000 years. Each type of boat—and its sails—was usually made from the most readily available materials in the area.

Ancient Egyptian Sailboat

Egyptian people used sailboats to move up and down the Nile River. They used square sails made from reeds that grew by the river. They used a long oar at the back of the boat to steer.

Bamboo Junk

In China, "junks" had sails made from bamboo. The sails were dipped in plant dye to make them yellow or red. Chinese sailors were the first to use rudders to steer their boats.

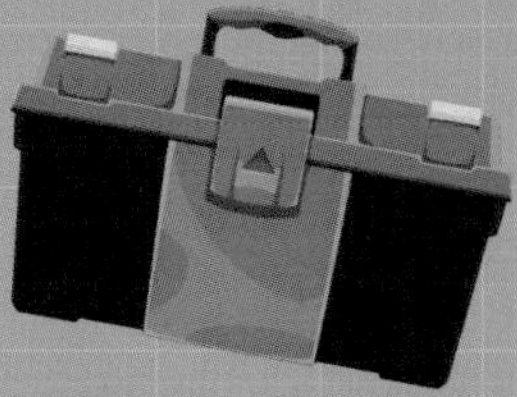

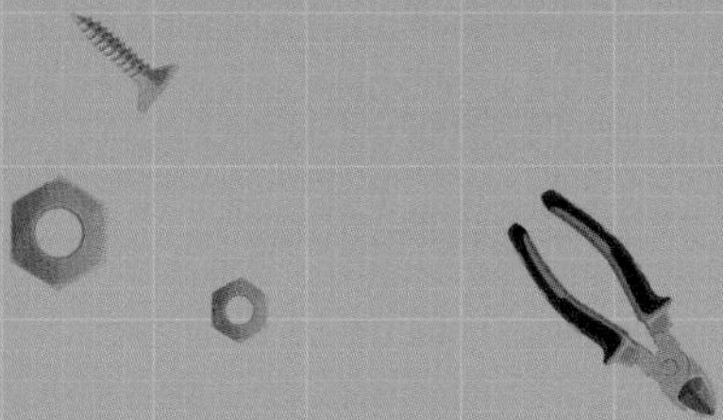

Umiak

In cold Arctic lands, boats called umiaks were made of wood. They were covered with sealskin or walrus skin to keep out water. Not all umiaks had sails. Those that did had sails made of animal skins or guts.

Viking Sailboat

Viking sailboats were built from wood, too. Sails on Viking boats were square. They were made of long threads of sheep's wool!

INDEX

ABOUT THE AUTHOR

Rebecca Stanborough lives and writes in Florida. She sees sailboats every day!